Geology

CONTENTS

Introduction 2
Curriculum Outline 3

SCIENCE 903
Teacher Notes 4
Answer Keys 5
Alternate Test & Key 12

SCIENCE 904
Teacher Notes 16
Answer Keys 17
Alternate Test & Key 22

SCIENCE 908
Teacher Notes 25
Answer Keys 27
Alternate Test & Key 32

SCIENCE 705
Teacher Notes 35
Answer Keys 37
Alternate Test & Key 45

SCIENCE 707
Teacher Notes 48
Answer Keys 49
Alternate Test & Key 56

Alpha Omega Publications®

804 N. 2nd Ave. E., Rock Rapids, IA 51246-1759
© MCMXCVII by Alpha Omega Publications, Inc. All rights reserved.
LIFEPAC is a registered trademark of Alpha Omega Publications, Inc.

All trademarks and/or service marks referenced in this material are the property of their respective owners. Alpha Omega Publications, Inc. makes no claim of ownership to any trademarks and/or service marks other than their own and their affiliates', and makes no claim of affiliation to any companies whose trademarks may be listed in this material, other than their own.

Dear Instructor,

Thank you for your interest in electives using the LIFEPAC Select Series.

The courses in this series have been compiled by schools using Alpha Omega's LIFEPAC Curriculum. These courses are an excellent example of the flexibility of the LIFEPAC Curriculum for specialized teaching purposes.

The unique design of the worktext format has allowed instructors to mix and match LIFEPACs from several subjects or grades to create alternative courses for junior high and high school credit.

Suggested support items for this course are the 7th and 9th Grade Science Experiments videos, SD0701 and SD0901. The videos include presentations of many of the experiments in this course. Several of the experiments that require special equipment or materials are demonstrated on these videos. They can either be used for answering the questions of the lab report or as a demonstration of the procedure prior to performing the experiment. A notice is included with each experiment in the LIFEPAC where the video is available.

These courses work particularly well as unit studies, as supplementary electives, or for meeting various school and state requirements. Another benefit of the courses—and any LIFEPAC subject, for that matter—is the ability to use them with any curriculum, at any time during the year, for any of several purposes:

- Elective Courses
- Make-up Courses
- Substitution Courses
- Unit Studies
- Summer School Courses
- Remedial Courses
- Multi-level Teaching
- Thematic Studies

Course Titles	*Suggested Credits*
Astronomy (Jr. High and above)	$\frac{1}{2}$ credit
Composition	$\frac{1}{2}$ credit
Geography	$\frac{1}{2}$ credit
Geology	$\frac{1}{2}$ credit
Life of Christ (Jr. High and above)	$\frac{1}{2}$ credit
Life Science	$\frac{1}{2}$ credit
Mankind: Anthropology and Sociology	$\frac{1}{2}$ credit

Geology

High School Level (1/2 credit)

Physical Geology
Science LIFEPAC 903

Earth Structures
- Shape
- Rocks
- Layers
- Igneous Structures
- Mountains

Earth Changes
- Weathering
- Erosion and Sedimentation

Earth Movements
- Isostasy
- Folding
- Folding
- Plate Tectonics

Historical Geology
Science LIFEPAC 904

An Observational Science
- The Science
- Sedimentary Rock
- Fossils
- Crustal Changes

Measuring Time
- Relative Time
- Absolute Time

Oceanography
Science LIFEPAC 908

History of Oceanography
- Chronology of Oceanography
- Techniques for Investigation
- Major Discoveries
- Submersible Research

Geology of the Ocean
- Geological Structure
- Results of Profiling
- Turbidity and Sedimentation
- World's System of Currents

Biological, Chemical, and Physical Properties
- Graphs
- Charts
- Other Resource Material

The Atmosphere
Science LIFEPAC 705

Structure of the Atmosphere
- Gases
- Layers
- Solar Effects
- Influences on Life
- Changes

Natural Cycles
- Water Cycle
- Carbon-Oxygen Cycle
- Nitrogen Cycle

Pollution
- Types of Pollutants
- Effects in Life
- Our Responsibility

Climate
Science LIFEPAC 707

Climate: General
- Weather
- Climate and Weather
- Parts of Climate

Climate: Worldwide
- Factors Affecting Climate
- General Distribution of Climate

Climate: Regional
- Classification of Climate
- Distribution of Climate Types

Climate: Local
- How Climate Affects People
- People, Communities, and Climate

Science 903 Teacher Notes

Materials Needed for LIFEPAC

Required:
pencil, desk or small table,
long sheet of paper, string, brick
double-pan balance,
set of metric masses, string,
various rocks, beaker

Suggested:
9th Grade Science Experiments Video
reference materials in either book or
online formats

Additional Learning Activities

Section 1: Earth Structures
1. Organize a field trip to a local rock or mineralogy museum.
2. With a classmate gather several rock samples. Your teacher may have some you can use. Check each sample for color and weight. See if you can determine whether the rocks are igneous, sedimentary, or metamorphic.
3. With a classmate find the directions in a library science reference book for making a "volcano." Construct one to show your class.
4. In the library science resource books look up *minerals*. Write a one-page report on minerals and how they relate to rocks.

Section 2: Earth Changes
1. Demonstrate the effect of sedimentation by stirring dirt into a beaker full of water. Set the beaker aside. Lead a class discussion on how sedimentation takes place in the oceans and lakes.
2. With a classmate demonstrate the effects of mechanical weathering. Heat a piece of glass tubing over the flame of a Bunsen burner. Be sure to use forceps. When the tubing is very hot, plunge it into cold water. What happens?
3. Demonstrate another effect of mechanical weathering with a friend. Fill a glass bottle with water. Wrap the bottle with a towel and set it in the freezer overnight. Carefully check the bottle. What happened?
4. Pour a mixture of small gravel, sand, and clay into a jar of water. Stir vigorously and let stand. Check the jar later to see what material has settled to the bottom first. Were layers formed?

Section 3: Earth Movements
1. With a friend trace the continents of the world on a map. Then cut each continent out. See if you can fit the "puzzle" pieces together.
2. Download and print out a topographical map of your area from the United States Geological Survey website.
3. Read a book or reliable online article on the drifting continents. Write a one-page report on the book. Be sure to include your scientific opinion of what is really happening to the continents. Make sure your opinion is reasonable based on your research.

Science 903 Answer Key

SECTION ONE

1.1 Eratosthenes
1.2 Ptolemy
1.3 sphere
1.4 During a lunar eclipse, the earth's shadow on the moon is circular.
1.5 The width was one-eighth of a minute of latitude.
1.6 true
1.7 true
1.8 true
1.9 igneous
1.10 lava
1.11 magma
1.12 Either order:
    a. tuff
    b. volcanic ash
1.13 b–c; either order:
    a. quartz
    b. feldspar
    c. mica
1.14 basalt
1.15 a. cooled from magma or lava
    b. laid in place by moving water, ice, or wind
    c. put under pressure, or under heat and pressure
1.16 a. sandstone
    b. conglomerate
    c. breccia
    d. siltstone or shale
1.17 a. slate or schist
    b. marble
    c. quartzite
1.18 true
1.19 false
1.20 true
1.21 true
1.22 false
1.23 false
1.24 c
1.25 a
1.26 c
1.27 b
1.28 hydrosphere
1.29 basalt
1.30 crust
1.31 mantle
1.32 asthenosphere
1.33 The gravitational pull of the earth is greater than the weight of surface rock.
1.34 Either order:
    a. earth's magnetic field
    b. meteorite composition
1.35 Either order:
    a. compression waves
    b. sideways shaking motions (vibrations)
1.36 Sideways vibrations do not move through liquids; sideways shock waves get lost at 2,900 kilometers.
1.37 a. The line is straight down the paper.
    b. The line is wavy and jagged rather than straight and smooth.
    c. The line has big notches (jagged marks) and these get smaller down to a straight even line.
1.38 true
1.39 true
1.40 false
1.41 false
1.42 false
1.43 c
1.44 b
1.45 Either order:
    a. silicon
    b. oxygen
1.46 size
1.47 pressure
1.48 intrusive
1.49 sill
1.50 joints
1.51 shield
1.52 Either order:
    a. silicon
    b. oxygen

Science 903 Answer Key

1.53 Any order:
   a. potassium
   b. aluminum
   c. sodium
   d. magnesium
   e. calcium or iron
1.54 Slow cooling allows time for molecules to move and come in contact with other similar molecules.
1.55 teacher check
1.56 false
1.57 true
1.58 true
1.59 true
1.60 false
1.61 a
1.62 c
1.63 Lava flows flow out of a fissure. A volcano is the result of lava finding a small place where the ground is weak.
1.64 huge waves of ocean water or dust in the sky

1.65 teacher check
1.66 false
1.67 true
1.68 false
1.69 scarp
1.70 magma
1.71 mesas
1.72 erosion
1.73 Either order:
   a. along zones of weakness
   b. over "hot spots"
1.74 Any order:
   a. shields of lava flows
   b. cinder cones
   c. combinations of a and b
1.75 Any order:
   a. volcanoes
   b. folded mountains
   c. fault-block
   d. domes
   e. erosional remnants

SECTION TWO

2.1 true
2.2 false
2.3 true
2.4 a
2.5 b
2.6 They can be turned into rock (stone).
2.7 Either order:
   a. chemical
   b. mechanical
2.8 Either order:
   a. plant roots
   b. decaying plants
2.9 exfoliation
2.10 talus
2.11 clay
2.12 Any order:
   a. on the floor plain of a river
   b. in a lake
   c. at the mouth of a river
   d. on a sandbar
   e. in a sand dune

2.13 Any order:
   a. wind
   b. water
   c. ice
2.14 Either order:
   a. dissolves minerals
   b. freezes and expands
2.15 topsoil – decayed vegetation
   subsoil – weathered rock
   regolith – partly weathered rock
   bedrock – unweathered rock
2.16 true
2.17 false
2.18 false
2.19 true
2.20 false
2.21 true
2.22 true
2.23 true
2.24 c
2.25 a

Science 903 Answer Key

2.26 d
2.27 d
2.28 a
2.29 c
2.30 d
2.31 b
2.32 a
2.33 headward erosion
2.34 alluvial fan
2.35 playa
2.36 flood plain
2.37 oxbow
2.38 old
2.39 desert
2.40 loess
2.41 plucking
2.42 striated
2.43 till
2.44 Any order:
    a. Desert sand is more uniform in grain size.
    b. Beach sand contains more silt.
    c. Beach sand may contain marine fossils.
2.45 A valley glacier is confined by ridges and flows downhill. A continental glacier covers areas hundreds of miles wide.
2.46 Any order:
    a. erratic boulders
    b. eskers
    c. moraines
    d. scratches (striations)
    e. drumlins
2.47 false
2.48 true
2.49 true
2.50 false
2.51 true
2.52 false
2.53 false
2.54 true
2.55 c
2.56 c
2.57 b
2.58 a
2.59 a

2.60 distributary
2.61 estuary
2.62 continental shelf
2.63 oolites
2.64 bacteria
2.65 ooze
2.66 reef
2.67 varves
2.68 water table
2.69 cavern
2.70 a. the rocks are broken into smaller pieces
    b. the rocks are worn smooth
2.71 Storms remove sand, leaving cobbles. Storms can destroy a beach.
2.72 Either order:
    a. nearby beach cliffs
    b. river – transported sand from hills and mountains
2.73 Nutrients (minerals) area available to grow algae which is eaten by fish and other sea life.
2.74 Rivers on land and ocean currents do not have enough energy to transport coarse material beyond the continental shelf.
2.75 lake – bog – solid ground (by infilling) or lake – river – (by destruction of the dam)
2.76 Any order:
    a. lake
    b. river
    c. spring

Science 903 Answer Key

SECTION THREE

3.1 true
3.2 true
3.3 true
3.4 true
3.5 true
3.6 true
3.7 false
3.8 a
3.9 c
3.10 density
3.11 fjords
3.12 isostasy
3.13 under high pressure
3.14 evaporation of seawater
3.15 salt domes
3.16 $\text{density} = \dfrac{\text{mass}}{\text{volume}} = \dfrac{4 \text{ grams}}{2 \text{ cm}^3}$

$\dfrac{2 \text{ grams}}{\text{cm}^3}$

3.17 2 (no units)
3.18 Plastic means that the rock will have a new shape or that it will flow due to heat and/or pressure.
3.19 A floating object is held up by a force equal to the weight (mass) of the substance displaced. For example, if a boat weighs 2,000 grams out of water and 1,000 in water, the boat in the water takes up the space of 1,000 grams of water.
3.20 The ice sheet is heavy enough to push the middle of the Antarctic continent down below sea level.
3.21 Either order:
   a. mining of salt
   b. as petroleum traps
3.22 c
3.23 d
3.24 geosynclines

3.25 isostasy
3.26 plateau
3.27 Either order:
   a. a former mountain range
   b. a more widespread source
3.28 b
3.29 a
3.30 false
3.31 true
3.32 true
3.33 true
3.34 Africa
3.35 Gondwanaland
3.36 rift valleys
3.37 Either order;
   a. Volcanoes
   b. earthquakes
3.38 folded
3.39 Mediterranean
3.40 Any order:
   a. Africa
   b. India
   c. Australia
   d. Antarctica
   e. South America
3.41 extends from Iceland south to the tip of Africa, then eastward into the Indian Ocean. Another ridge lies in the eastern Pacific.
3.42 plastically flowing rock
3.43 Any order:
   a. volcanoes and earthquakes that circle the Pacific
   b. mountain areas where plates descend
   c. mid-ocean ridges
   d. ocean trenches

Science 903 Self Test Key

Self Test 1

1.01	e		1.021	b
1.02	k		1.022	b
1.03	i		1.023	d
1.04	g		1.024	a
1.05	a		1.025	c
1.06	j		1.026	c
1.07	b		1.027	d
1.08	h		1.028	b
1.09	f		1.029	a
1.010	d		1.030	d
1.011	false		1.031	volcanic shield
1.012	false		1.032	weakness
1.013	true		1.033	sea mounts or guyots
1.014	false		1.034	batholith
1.015	true		1.035	granite
1.016	true		1.036	meteorites
1.017	false		1.037	core
1.018	false		1.038	nitrogen
1.019	true		1.039	metamorphic
1.020	false		1.040	sandstone

Self Test 2

2.01	f		2.020	true
2.02	d		2.021	c
2.03	a		2.022	a
2.04	g		2.023	b
2.05	b		2.024	a
2.06	j		2.025	d
2.07	e		2.026	d
2.08	k		2.027	a
2.09	h		2.028	b
2.10	i		2.029	b
2.011	false		2.030	b
2.012	true		2.031	flood plain
2.013	false		2.032	meanders
2.014	true		2.033	dry or desert
2.015	true		2.034	erratic(s)
2.016	true		2.035	turbidity
2.017	false		2.036	shelf
2.018	false		2.037	river
2.019	false		2.038	calving

2.039 water vapor
2.040 sedimentary
2.041 Whether rock material is transported and then deposited. The deposited rock fragments can be back into rock.

Self Test 3

3.01   e
3.02   a
3.03   h
3.04   g
3.05   j
3.06   i
3.07   k
3.08   d
3.09   c
3.010 b
3.011 false
3.012 false
3.013 true
3.014 true
3.015 true
3.016 true
3.017 false
3.018 false
3.019 true
3.020 true
3.021 a
3.022 c
3.023 c
3.024 a
3.025 a
3.026 d
3.027 a
3.029 b
3.030 d
3.031 density
3.032 sedimentary
3.033 plateau
3.034 volcanoes
3.035 southern
3.036 igneous
3.037 magma
3.038 lava
3.039 oxbow
3.040 weathering
3.041 density $= \dfrac{\text{mass}}{\text{volume}} = \dfrac{3.6\text{g}}{1.2\,\text{cm}^3} = 3\,\dfrac{\text{g}}{\text{cm}^3}$

3.042 Sp. gr. $= \dfrac{\text{mass in air}}{\text{mass in air} - \text{mass in water}} = \dfrac{160\text{g}}{160\text{g} - 100\text{g}} = \dfrac{160\text{g}}{60\text{g}} = 2.67$

Science 903
LIFEPAC Test

1. c
2. e
3. a
4. b
5. d
6. c
7. b
8. a
9. a
10. c
11. c
12. b
13. c
14. a
15. a
16. d
17. f
18. a
19. b
20. c
21. false
22. true
23. false
24. true
25. false
26. true
27. true
28. true
29. false
30. false
31. true
32. true
33. true
34. false
35. true
36. false
37. false
38. false
39. false
40. false
41. c
42. d
43. a
44. b
45. c
46. a
47. c
48. c
49. a
50. b

Science 903 Alternate Test

Name _____

Write the letter for the correct answer on each line (each answer, 2 points).

1. Granite is _____ .
   a. metamorphic   b. sedimentary   c. igneous

2. Sandstone is _____ .
   a. metamorphic   b. sedimentary   c. igneous

3. Basalt is _____ .
   a. metamorphic   b. sedimentary   c. igneous

4. Marble is _____ .
   a. metamorphic   b. sedimentary   c. igneous

5. Shale is _____ .
   a. metamorphic   b. sedimentary   c. igneous

6. Feldspar and mica converting to clay minerals is an example of _____ .
   a. chemical weathering   c. erosion
   b. physical weathering   d. ooze

7. Domed mountains are pushed upward by masses of _____ .
   a. lava   c. gases
   b. magma   d. sedimentation

8. Volcanic debris does *not* include _____ .
   a. lava   c. tuff
   b. magma   d. cinder

9. Gases in a volcano are primarily _____ .
   a. water vapor   c. sulfur dioxide
   b. carbon dioxide   d. nitrogen

10. Steep-sided glaciated valleys are called _____ .
    a. domes   c. graded beds
    b. basins   d. fjords

Science 903 Alternate Test

Complete these sentences (each answer, 3 points).

11. To tell whether an igneous rock cooled slowly or rapidly, check the size of the _____ .
12. Magma that penetrates buried rocks, but does not reach the surface, is called _____ rock.
13. Magma that erupts from a volcano is called _____ .
14. Bends in a river crossing a lowland are called _____ .
15. The mass of a unit volume of material is called _____ .

Answer *true* or *false* (each answer, 1 point).

16. _____ Sir Isaac Newton reasoned that the earth bulges slightly at the center.
17. _____ The three main categories of rock are igneous, sedimentary, and metamorphic.
18. _____ Siltstone is an example of a metamorphic rock.
19. _____ The middle part of the earth's core is liquid.
20. _____ Earthquakes are measured by a seismograph.
21. _____ Volcanoes that have fast flowing liquids will be tall and thin.
22. _____ Regolith is partly weathered rock.
23. _____ A branch of a river is called a distributary.
24. _____ The density of an object compared to the density of air is called specific gravity.
25. _____ Glaciers that travel between ridges are called valley glaciers.
26. _____ Lava may be called ashes, cinders, or bombs.
27. _____ Major mountain types include fold, fault, dome, and erosional remnants.
28. _____ Mechanical weathering always involves a chemical change.
29. _____ A lake that evaporates is called a playa.
30. _____ A teardrop-shaped moraine is called a drumlin.

Science 903 Alternate Test

Match these items (each answer, 2 points).

31.	_____ designer of the universe	a.	weathering
32.	_____ first to calculate circumference of the earth	b.	God
		c.	seamount
33.	_____ rocks changed by heat and pressure	d.	old river
		e.	moraine
34.	_____ Second layer of the earth	f.	Eratosthenes
35.	_____ former volcanoes now under the sea	g.	anticline
		h.	mesa
36.	_____ flat-topped hill	i.	metamorphic
37.	_____ chemical and mechanical process that breaks up rocks	j.	avalanche
		k.	mantle
38.	_____ crosses a nearly flat flood plain		
39.	_____ snow falling down a mountain side		
40.	_____ top of a fold		

Complete these activities (each answer, 5 points).

41. Explain the difference between sedimentary and igneous rocks.
    _____
    _____

42. Explain the difference between mechanical weathering and chemical weathering.
    _____
    _____
    _____

64 / 80

Date _____

Score _____

Science 903 Alternate Test Key

1. c
2. b
3. c
4. a
5. b
6. a
7. b
8. b
9. a
10. d
11. crystals
12. intrusive
13. lava
14. meanders
15. density
16. true
17. true
18. false
19. false
20. true
21. false
22. true
23. true
24. false
25. true
26. true
27. true
28. false
29. true
30. true
31. b
32. f
33. i
34. k
35. c
36. h
37. a
38. d
39. j
40. g
41. Example:
    Sedimentary rocks are formed in layers under the sea; igneous rocks are fire-formed.
42. Example:
    Mechanical weathering involves a physical change, and chemical weathering involves a chemical change.

Science 904 Teacher Notes

Materials Needed for LIFEPAC

Required:

Suggested:
felt tip pen, one dozen plastic sandwich bags with ties, pen or pencil, two paper grocery bags, clipboard, twenty 3" x 5" cards reference materials in either book or online formats

Additional Learning Activities

Section 1: An Observational Science

1. With a classmate visit the section on geology at your local science museum.
2. Do some research on how the earth's history has been divided into different periods. Make a chart showing the geologic time scale used in this country.
3. Write a two-page report on how the work with fossils done by Baron Cuvier and William Smith has helped geologists determine the age of different rock layers.

Section 2: Measuring Time

1. With your class use a world map or globe to find the location of the areas discussed in Section 2.
2. With a friend make a sundial and use it to tell solar time. Most encyclopedias or general science books have directions for making sundials.
3. With a classmate study a map that has time zones. When it is four o'clock where you live what time is it in San Francisco? New York? Honolulu?
4. Write a two-page science fiction paper telling what you think historical geologists will find in the year 2037. What methods will they use to determine what life was like in 1996?
5. Make a wall chart illustrating the difference between relative time and absolute time.

Science 904 Answer Key

SECTION ONE

1.1 true
1.2 false
1.3 b
1.4 b
1.5 Neptunists
1.6 Plutonists
1.7 France
1.8 conclusion
1.9 Basalt was traced to the volcanic vent from which it erupted.
1.10 a. time
 b. materials
1.11 Geology is an observational science because a geologist cannot do his work in a laboratory. He must depend on clues in the rocks.
1.12 A sedimentary bed could be recognized by fossils it contains.
1.13 teacher check
1.14 teacher check
1.15 teacher check
1.16 false
1.17 false
1.18 true
1.19 true
1.20 false
1.21 b
1.22 d
1.23 c
1.24 b
1.25 c
1.26 high
1.27 surface or crust
1.28 fossils
1.29 granite
1.30 clay
1.31 lithification
1.32 clastic (detrital)
1.33 formation
1.34 member
1.35 group
1.36 Organisms are buried in sedimentary rock and fossilized. Sedimentary rocks contain fossils.
1.37 Limestone is deposited in oceans. Something happened to raise a continental shelf to the height of a mountaintop.
1.38 A break in sedimentation represents a time during which sediment was not deposited.
1.39 Sedimentary rocks form from detritus (sediment) that has been weathered from first-generation granites.
1.40 a. water chemically weathers the minerals
 b. the stream transports the sediment
 c. the stream winnows (separates) the different kinds of sediment
1.41 Detritus is particles of pre-existing rocks; limestone comes from organisms.
1.42 a. reducing the volume, increasing the density – squeezing the sediment grains together.
 b. the growing together of grains that are in contact.
 c. the bonding of grains by salts (cement) precipitated from the sea water.
1.43 a. silt
 b. algal structures, coral frags, seashells, sponge spines
 c. gravel
 d. limestone
 e. mudstone
 f. conglomerate
 g. sandstone
 h. limestone
1.44 a. sand
 b. mud
 c. gravel
1.45 a. group
 b. formation
 c. member
 d. bed
1.46 a. It must have shell, bones, etc.
 b. It must be buried quickly

Science 904 Answer Key

1.47 a. oxygen
      b. bacteria
      c. scavengers
1.48 a. petrifaction-replacement
      b. total removal, leaving a mold.
      c. filling of the mold producing a cast.
      d. removal of liquids and solids, leaving carbon-distillation
      e. freezing in regions of constant cold. Also mummification in peat bogs and oil seeps
1.49 Mud does not permit the flow-through of water that carries oxygen.
1.50 a. Good: sediment for rapid burial
      b. Bad: scavengers – no burial
      c. Good: low oxygen content
1.51 teacher check
1.52 teacher check
1.53 a. the search for oil
      b. the search for minerals
      c. the planning of bridges, tunnels, and dams
1.54 true
1.55 false
1.56 false
1.57 true
1.58 false
1.59 true
1.60 true
1.61 true
1.62 true
1.63 false
1.64 c
1.65 a
1.66 c
1.67 c
1.68 c
1.69 b
1.70 a. calcite
      b. silica
1.71 poor
1.72 mold
1.73 turned to rock
1.74 trace
1.75 paleontology
1.76 orogeny

SECTION TWO

2.1 false
2.2 true
2.3 true
2.4 false
2.5 false
2.6 true
2.7 false
2.8 d
2.9 c
2.10 relative
2.11 correlating
2.12 Superposition
2.13 a. turbidites
      b. thrust faults
2.14 dike
2.15 unconformity
2.16 All members of a graded bed formed by a turbidity current are deposited simultaneously.
2.17 In water, fine sediment settles more slowly than coarse sediment.
2.18 teacher check
2.19 a. campsite in three to twelve meters of water off La Jolla, California.
      b. the seacoast town of Limani Chersoniso, built by the Romans inland.
      c. Venice, Italy, sinking
      d. Long Beach, California
      e. temple of Jupiter Serapis near Pozzuoli, Italy, which shows signs of rise and fall, or a beach on the Baltic Sea, which is 500 meters above sea level
2.20 dike
2.21 M
2.22 orogeny
2.23 unconformity

Science 904 Answer Key

2.24	b		2.44	varve
2.25	d		2.45	organic
2.26	d		2.46	Either order:
2.27	c			a. Sequoia
2.28	false			b. bristlecone pine
2.29	true		2.47	Genesis
2.30	true		2.48	relative
2.31	true		2.49	absolute
2.32	false		2.50	absolute
2.33	false		2.51	relative
2.34	true		2.52	relative
2.35	d		2.53	absolute
2.36	b		2.54	absolute
2.37	d		2.55	relative
2.38	d		2.56	absolute
2.39	b		2.57	5
2.40	c		2.58	2
2.41	c		2.59	5
2.42	a. temperature		2.60	3
	b. food supply		2.61	Either order:
2.43	Any order:			a. k
	a. light			b. y
	b. temperature			
	c. precipitation			

Science 904 Self Test Key

Self Test 1

1.01	e	1.022	true
1.02	d	1.023	true
1.03	f	1.024	true
1.04	f	1.025	true
1.05	j	1.026	a
1.06	k	1.027	c
1.07	h	1.028	a
1.08	b	1.029	b
1.09	c	1.030	b
1.010	i	1.031	d
1.011	true	1.032	a
1.012	false	1.033	b
1.013	true	1.034	a
1.014	true	1.035	b
1.015	false		
1.016	true		
1.017	false		
1.018	false		
1.019	false		
1.020	false		
1.021	true		

1.036 Because sedimentary rocks are made of fragments of other rocks.

1.037
 a. sandstone
 b. conglomerate
 c. siltstone
 d. claystone
 e. limestone

Self Test 2

2.01	d	2.021	false
2.02	e	2.022	true
2.03	g	2.023	true
2.04	h	2.024	true
2.05	a	2.025	false
2.06	b	2.026	clastic
2.07	c	2.027	tree rings
2.08	j	2.028	quartz
2.09	k	2.029	petrifaction
2.010	i	2.030	lithification
2.011	a		
2.012	a		
2.013	b		
2.014	c		
2.015	a		

2.031 Example: The old Roman inland town of Limani Chersonisos is now on the seacoast.

2.016	true	2.032	teacher check
2.017	false	2.033	U
2.018	false	2.034	D
2.019	true	2.035	A
2.020	false	2.036	unconformity
		2.037	Superposition
		2.038	thrust

Science 904
LIFEPAC Test

1. graded bed
2. chemical
3. mud
4. conglomerate
5. lithification
6. recrystallization
7. petrifaction
8. orogeny
9. paleontology
10. underwater
11. unconformity
12. thrust
13. Superposition
14. absolute
15. varves
16. c
17. a
18. c
19. a
20. c
21. a. group
    b. formation
    c. member
    d. bed
22. clay
23. silt
24. sand
25. sponge spines/coral fragments/seashells/algal structures
26. gravel

Science 904 Alternate Test

Name _____

Match these items (each answer, 2 points).
1. _____ turned to stone
2. _____ having a backbone
3. _____ the order in which rocks are placed one above the other
4. _____ science of fossils
5. _____ fragments, rubbish
6. _____ formed at the earth's surface
7. _____ in relation to others
8. _____ surface of erosion
9. _____ found in the sea
10. _____ gravel, sand, mud

a. relative
b. debris
c. unconformity
d. detrital sediments
e. marine
f. orogenic
g. vertebrate
h. sedimentary rocks
i. petrified
j. paleontology
k. superposition

Complete these sentences (each answer, 3 points).
11. Fossils that do not reveal the body form are called _____.
12. Precipitation of a binding material around grains in rocks is called _____.
13. The study of plants through the study of fossils is called _____.
14. Orogeny is the cause of _____.
15. A climatic calendar for the United States Southwest has been compiled from information gained from _____.
16. Organisms without backbones are called _____.
17. The group of scientists who believed that crustal rock arrived at the surface of the earth in molten state was called _____.
18. Physics, chemistry, and biology are called _____ sciences.
19. The parts of an organism that decay first are the _____ parts.
20. The idea that younger formations overlie older formations is the law of _____.

Science 904 Alternate Test

Complete these activities (each answer, 5 points).

21. Explain the difference between the Plutonists and the Neptunists.
   _____
   _____
   _____

22. State several reasons why sedimentary rocks are of interest to historical geologists. _____
   _____
   _____
   _____

48 / 60

Date_____

Score_____

Science 904 Alternate Test Key

1. i
2. g
3. k
4. j
5. b
6. h
7. a
8. c
9. e
10. d
11. trace fossils
12. cementation
13. paleobotany
14. crustal changes
15. tree rings
16. invertebrates
17. Plutonists
18. laboratory
19. fleshy
20. superposition
21. Example:
Plutonists believed that crustal rock arrived at the surface of the earth in molten state. Neptunists believed that all crustal rock was precipitated from an ocean.
22. Example:
Sedimentary rocks form at the earth's surface, are the burial ground for former life, preserve a record of life and environments, and represent the passage of time.

Science 908 Teacher Notes

Materials Needed for LIFEPAC
Required:

Suggested:
almanac
reference materials in either book
or online formats

Additional Learning Activities

Section 1: History of Oceanography
1. Research and make a large chart listing all of the important discoveries and innovations in oceanography.
2. Research and write a one-page report on the effect of the moon on tidal reaction.

Section 2: Geology of the Ocean
1. Have class construct a plaster of Paris model representing the major typographical features of the ocean floor, including such things as the mid-ocean ridges, rifts, deep trenches, seamounts, and continental shelves, all made to a reasonably accurate scale. Once the model is completed, it can be an excellent training aid to assist students in their understanding of the theories of sea-floor spreading and global tectonics.
2. Search the websites of the following groups for information on the geology of the ocean: Naval Oceanographic Office (Suitland, MD), U.S. Coast Guard and Smithsonian Institution (Washington, D.C.), and the National Ocean Survey (Rockville, MD). Also, there are a number of educational and nonprofit institutes—Scripps Institution of Oceanography (La Jolla, CA) and Woods Hole Oceanographic Institution (Woods Hole, MA); Texas A&M (College Station, TX); Oregon State University (Corvallis, OR); University of Rhode Island (Kingston, RI); University of Washington (Seattle, WA); Johns Hopkins (Baltimore, MD); and the University of Miami (Miami, FL).
3. With a friend make a chart showing the ocean currents.
4. Write a one-page report explaining the effect of sea-floor spreading in producing oceanic trenches.
5. Research the life and discoveries of V.W. Ekman. Explain the Ekman Spiral in your report.

Section 3: Biological, Chemical, and Physical Properties
1. Have the class obtain data available in various world almanacs concerning world production and consumption of petroleum. Also, have them obtain clippings of recent magazine and newspaper stories of noteworthy events which bear on the political, economic, and social considerations relative to the oil-energy situation. Encourage the class to analyze the cause and effect relationship regarding decisions by countries with regard to an energy crisis and how the science of oceanography may be the answer to an ever increasing problem.

Science 908 Teacher Notes

2. Assign a student the task of making a full report of fish meal production; another one on its role in the food chain process which occurs off the coast of Peru; and a third on its value as a source of protein compared to other marine life.
3. Research the life of Jacques Cousteau. Write a paper on his importance to the science of oceanography.
4. With a friend make a poster of the carbon dioxide cycle.
5. Write a one-page paper explaining what causes ocean waves.

Science 908 Answer Key

SECTION ONE

1.1 i
1.2 g
1.3 h
1.4 f
1.5 a
1.6 j
1.7 e
1.8 d
1.9 b
1.10 c
1.11 He ordered the production of Gulf Stream charts which enabled faster ocean crossing for the mails between England and the colonies.
1.12
a. Its 3-year voyage brought back so much data as to interest greatly the scientific community in oceanographic research.
b. covered over 140 million square miles of ocean in 3 years
c. Its chemist, Buchanan, was credited with founding chemical oceanography.
d. Its scientific reports provide 50 volumes of data on the seas.
1.13 Example:
Haber sought gold from seawater in 1925, which was unsuccessful. However, his data on salt nutrient distribution in relation to plankton proved very beneficial.
1.14 Radio Direction Finder
1.15 subbottom profiles
1.16 4,650 feet
1.17 The current drag is submerged beneath the surface and measures currents independently of the surface float which is affected by surface winds and currents.
1.18 d
1.19 a
1.20 f
1.21 e
1.22 b
1.23 c
1.24 mid-Atlantic ridge
1.25
a. Cousteau
b. *Conshelf Two*
1.26 Any order:
a. uranium
b. silver
c. gold
1.27 e
1.28 g
1.29 a
1.30 f
1.31 c
1.32 b

SECTION TWO

2.1 Any order:
a. continental drift
b. sea-floor spreading
c. global tectonics
2.2 A transmitter sends a sound wave that is reflected to a receiver. From the time required for the return, the depth can be calculated; and a profile of the ocean floor can be derived.
2.3 continuous seismic profiling
2.4 seismic refraction
2.5 coring
2.6
a. hundred cores
b. 1,000
2.7 ecogram
2.8 c
2.9 f
2.10 e

Science 908 Answer Key

2.11 b
2.12 d
2.13 Europe
2.14 pressures
2.15 a. pressure
      b. 14,000
2.16 false
2.17 true
2.18 false
2.19 true
2.20 Sediments are classified according to their size, chemical composition and place of deposit.

2.21 Sea-floor sediments are increasingly thick away from the mid-ocean ridges because they have had more time to accumulate sediment.
2.22 Because they are near the western boundaries of their ocean systems.
2.23 tuna gear was observed to be moving eastward below westward moving surface currents
2.24 friction between thin water layers
2.25 d
2.26 c
2.27 a

SECTION THREE

3.1   false
3.2   true
3.3   true
3.4   false
3.5   true
3.6   copepod
3.7   its ease of capture
3.8   zooplankton
3.9   trawl net
3.10 e
3.11 a
3.12 b
3.13 f
3.14 d
3.15 c
3.16 d
3.17 a
3.18 c
3.19 b
3.20 a. hydrogen
      b. oxygen
      c. 96

Any order:
      d. chlorine
      e. 2
      f. sodium
      g. 1
      h. magnesium
      i. 0.1
3.21 c
3.22 e
3.23 d
3.24 a
3.25 b
3.26 tsunamis
3.27 gravity
3.28 Arctic Ocean
3.29 France
3.30 the petroleum embargo
3.31 16 percent

Science 908 Self Test Key

SELF TEST 1

1.01 significant wave
1.02 Either order:
    a. *Nautilus*
    b. *Skate*
1.03 fractures
1.04 from the bottom
1.05 electromagnetic radiation
1.06 g
1.07 e
1.08 d
1.09 b
1.010 h
1.011 i
1.012 c
1.013 a
1.014 j
1.015 f
1.016 Either order:
    a. Edward Forbes
    b. Matthew Maury
1.017 Any order:
    a. Alexander Agassiz
    b. W.C. McIntosh
    c. Frank Buckland
1.018 a. Sealab I
    b. DSRV-1
    c. Trieste
    d. Sealab II

SELF TEST 2

2.01 false
2.02 true
2.03 false
2.04 true
2.05 true
2.06 Palomares, Spain
2.07 17,000 feet
2.08 Either order:
    a. mid-ocean ridges
    b. ocean trenches
2.09 Pacific Plate
2.010 turbidity currents
2.011 d
2.012 h
2.013 a
2.014 j
2.015 b
2.016 e
2.017 c
2.018 i
2.019 f
2.020 g
2.021 28.75 mph
2.022 62.5 in^3
2.023 100 million years

SELF TEST 3

3.01 This order only:
   a. China
   b. Peru
   c. Indonesia
   d. United States
   e. Japan
   f. India

3.02 Any order:
   a. amount of dissolved material (chlorine content)
   b. amount of evaporation
   c. amount of precipitation
   d. amount of fresh water introduced by rivers
   e. the mixing of currents

3.03 false

3.04 true

3.05 true

3.06 false

3.07 true

3.08 ocean-atmosphere exchange process

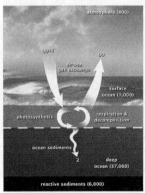

3.09 tidal wave bulge effect

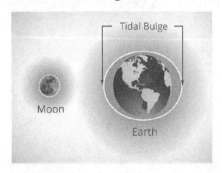

3.010 wave formation

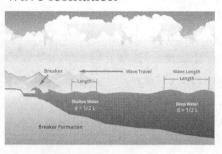

3.011 Ekman spiral

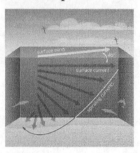

3.012 0.3 knots
3.013 f
3.014 j
3.015 l
3.016 a
3.017 b
3.018 m
3.019 c
3.020 d
3.021 e
3.022 g
3.023 h
3.024 i
3.025 k
3.026 n

Science 908
LIFEPAC Test

1. b
2. a
3. c
4. b
5. a

6. true

7. false
8. true
9. false
10. false
11. true
12. false
13. true
14. true
15. false
16. a unit of measure equal to six feet, used mostly in measuring depth in water
17. the process by which simple sugars and starches are produced from carbon dioxide and water by living plant cells with the aid of chlorophyll and in the presence of light
18. small animal and plant organisms that float or drift in water which serve as an important source of food for larger fish
19. the process used to produce fresh water from saline, usually by distillation, but also to a small degree by crystallization
20. a pattern of feeding relationships among organism

21. Victor Hansen
22. C.G.J. Petersen
23. Cousteau and Gagnan
24. seismic refraction
25. Any order:
    a. subterranean earthquake
    b. turbidity currents
    c. volcanic eruptions
26. DSRV-1
27. clockwise
28. 8 feet

Science 908 Alternate Test

Name _____

Match these items (each answer, 2 points).
1. _____ Scott
2. _____ laser studies
3. _____ continental slopes
4. _____ diving saucer
5. _____ echo sounding
6. _____ East Indian Rift
7. _____ sedimentation process
8. _____ Woods Hole
9. _____ marine life extracted
10. _____ floating net-rafts
11. _____ underwater pump and electric lights
12. _____ upwelling
13. _____ 5°F increase world-wide
14. _____ krypton
15. _____ higher degree of salinity

a. Conshelf Two experiment
b. estimated to have occurred 20 million years ago
c. releases 15,000 bottles a year for current measurement
d. about 2/3 for human consumption
e. largely used by Japanese
f. produces about 99 percent of fish available
g. biology of Antarctic whales
h. enough to raise ocean level by 250 feet
i. drop off 100–500 feet per mile
j. Ice Age
k. Cromwell Current
l. inert gas
m. echogram
n. lowers freezing point of water
o. Russian development for fish extraction

Complete these statement (each answer, 3 points).
16. The most critical skill needed by mariners sailing across the sea was _____ .
17. A device used for attracting fish by use of sound reproduction is a _____ .
18. Sir Isaac Newton's law of gravitation provided a mathematical basis for _____ .
19. Whales are classified as _____ consumers in the ocean food web.
20. Laterally moving convection currents are thought to be a major reason for creating _____ .

Science 908 Alternate Test

Write the letter for the correct answer on each line (each answer, 2 points).
21. The French tidal-power station has a capacity of _____ .
    a. 35,000 horsepower   b. 240 megawatts   c. tsunami
22. It is estimated that the world's petroleum reserves amount to _____ .
    a. 700 billion barrels   b. unlimited amounts   c. 1.2 trillion barrels
23. Confirmation of the ocean's deepest trench was achieved by the _____ .
    a. HMSM Challenger   b. Glomar Challenger   c. Trieste
24. The United States ranks fourth among fish-catch nations, but its fish consumption is _____ of the world's average.
    a. double   b. triple   c. quadruple
25. During the massive turbidity current off Newfoundland, transatlantic cables were found snapped as much as _____ away.
    a. 100 miles   b. 50 miles   c. 300 miles
26. The main sponsor of the Conshelf Two experiment was _____ .
    a. Ecole Polytechnique
    b. French national petroleum office
    c. Woods Hole

46 / 57

Date_____

Score_____

Science 908 Alternate Test Key

1. g
2. k
3. i
4. a
5. m
6. b
7. j
8. c
9. d
10. e
11. o
12. f
13. h
14. l
15. n
16. navigation
17. recording
18. tidal theory
19. secondary
20. ocean trenches
21. b
22. a
23. c
24. a
25. c
26. b

Science 705 Teacher Notes

Materials Needed for LIFEPAC

Required:
shoe boxes
thermometers
sheet of clear plastic or glass
cellophane wrap
stopwatch; any watch that has a
second hand
pan; 20 to 30 cm diameter by 5 cm deep
metric ruler

Suggested:
7th Grade Science Experiments
Video
reference materals in either book
or online formats

ADDITIONAL LEARNING ACTIVITIES

Section 1: Structure of the Atmosphere

1. Select one of the gases from Figure 2 Section 1 of the LIFEPAC to write a one-page report about. Use an encyclopedia, reliable online resources, or another reference book to find information for the report.
2. Using an encyclopedia or reliable online resources compare the summer and winter temperatures of several states and countries to determine whether they gain or lose more radiation. Prepare a chart or table to show the results of your study.

Section 2: Natural Cycles

1. Demonstrate the carbon-oxygen cycle by sealing a snail and an aquarium plant such as *Anacharis* (elodea) in a test tube or baby food jar. (Both organisms may be obtained at any pet or variety store that has tropical fish.)
2. Illustrate the principles involved in the water cycle with a distillation apparatus. You will need a flask that has been fitted with a one-hole stopper, a U-shaped piece of glass tubing, and a test tube resting in a beaker of ice water.

Science 705 Teacher Notes

3. Obtain the roots of alfalfa, clover, peanut, or other legume plants. Study the swellings or nodules on these roots. They contain thousands of nitrogen-fixing bacteria.
4. Students may study the effects of nitrogen-fixing bacteria on plant growth in the classroom. Sterilize four small flowerpots of soil in an oven (320° to 375° for at least one hour). When the soil has cooled, add nitrogen-fixing bacteria (purchased from a seed company) to the soil in two of the flowerpots. Soak clover seeds in water for several hours and add equal quantities of the seeds to each of the four pots. Observe the growth for several weeks. In which pots do the students notice the healthiest clover plants? Transpiration may be observed in a growing plant by covering the soil in the pot with plastic wrap and placing the plant under a large glass jar in bright sunlight. Within fifteen minutes a film of water will form on the inside of the jar. (The plastic wrap is necessary to prevent the evaporation of soil water.)
5. Write a report on seeding clouds with dry ice or iodized salts to produce rain.

Section 3: Pollution
1. The effects of sulfur oxides on a living plant may be demonstrated by melting a small amount of powdered sulfur (available at drug stores) in a closed container with a plant. Geraniums are especially good for this demonstration. For a control, place a second geranium in a closed container but do not melt the sulfur in the control. After a few days the effects of the sulfur gas will be dramatic as the plant turns brown and dies. Note: This experiment should only be conducted by the teacher. Inhaling large amounts of the gas is dangerous.
2. Prepare a bulletin board on some aspect of the air pollution problem. For instance, you might consider the causes of pollution, the results of pollution, or ways to control pollution as possible themes.
3. Study about the composition of glass, metal cans, and newspaper. Write a one-page report on what is saved when these items are recycled.
4. Clip articles about pollution from newspapers and magazines or printouts of online articles and make a pollution scrapbook.

Science 705 Answer Key

SECTION ONE

1.1 50%
1.2 nitrogen
1.3 oxygen
1.4 water vapor
1.5 Oxygen molecules consist of two oxygen atoms. Ozone molecules are made of three oxygen atoms.
1.6 30 to 60 km above the earth in the ozonosphere
1.7 a region where the concentration of ozone is high.
1.8 temperature
1.9 troposphere
1.10 stratosphere
1.11 mesosphere
1.12 thermosphere
1.13 ionosphere
1.14 b
1.15 c
1.16 a
1.17 b
1.18 a
1.19 troposphere means sphere of overturning
1.20 troposphere
1.21 The top of the stratosphere is warmer than the bottom. Therefore, the cold heavy air is on the bottom.
1.22 Gases in the troposphere, stratosphere, and mesosphere are uniformly mixed. The gases in the thermosphere are found in layers.
1.23 a
1.24 b
1.25 b
1.26 c
1.27 Both layers are warmed from below.
1.28 Gravity arranges the heaviest gases at the bottom and the lightest gases on top.
1.29 It is meaningless because the air is too thin to heat objects passing through it
1.30 a. It is formed when nitrogen and oxygen absorbed radiation to become ions.
   b. It is important because it reflects radio waves.
1.31 sun
1.32 a. 30
   b. 30
   c. 40
1.33 Either order:
   a. ionosphere
   b. ozonosphere
1.34 short-wave
1.35 Either order:
   a. carbon dioxide
   b. water vapor
1.36 the earth
1.37 night
1.38 a. equator
   b. poles
1.39 The bending of light rays in all directions by gas molecules.
1.40 scattering
1.41 The radiation released by the sun has a shorter wavelength than radiation released by the earth.

Science 705 Answer Key

1.42 It is the trapping of heat given off by earth after having been heated by the sun. Carbon dioxide and water vapor trap the heat.

1.43 The amount of radiation received by the earth equals the amount of heat lost to space.

1.44 The temperature rose faster in the covered box. (student answers may vary)

1.45 The temperature rose higher in the covered box. (student answers may vary)

1.46 The temperature dropped faster in the uncovered box. (student answer may vary)

1.47 Example:
The plastic or glass acted like carbon dioxide and water vapor in the atmosphere. Sunlight (short waves) can pass through the plastic or glass. Heat (long waves) in the box cannot escape. As a result the temperature rises. In the absence of sunlight the plastic keeps the heat in the box.

1.48 oxygen

1.49 Either order:
  a. water
  b. carbon dioxide

1.50 energy

1.51 Either order:
  a. food
  b. oxygen

1.52 ionosphere
1.53 ozonosphere
1.54 burns, cancer, or even death
1.55 thermosphere
1.56 craters
1.57 atmosphere

1.58 Either order:
  a. stone
  b. metal

1.59 true
1.60 false

1.61 The process by which oxygen combines with food to produce energy.

1.62 The process by which plants use chlorophyll and sunlight to combine carbon dioxide and water to make oxygen and food.

1.63 sunlight, chlorophyll, carbon dioxide, water

1.64 The skin is tanned by small amounts of radiation. Larger amounts can cause the skin to be thin and wrinkled and sometimes causes cancer.

1.65 Radiation shortens the life span.

1.66 bone marrow, sex organs, digestive system, blood vessels

1.67 Atmosphere moderates a planet's temperature: its temperature range is small.

1.68 solid objects moving through space at high speeds

1.69 Meteors (meteoroids in the atmosphere) burn up because they are heated by friction with air molecules.

1.70 Most meteoroids are the size of a grain of sand.

1.71 It is a meteoroid that reaches the earth's surface.

1.72 The atmosphere causes most meteoroids to burn up.

1.73 warming trend
1.74 cool
1.75 true

1.76 More carbon dioxide will trap more heat (the greenhouse effect) and raise temperatures on earth.

1.77 The burning of fossil fuels adds carbon dioxide to the atmosphere.

1.78 They give off carbon dioxide and water vapor.

1.79 Example:
Temperatures could become warmer over the entire earth. Warmer temperatures might melt the ice caps and glaciers, raising the ocean levels.

Science 705 Answer Key

SECTION TWO

2.1 c

2.2 e

2.3 b

2.4 a

2.5 d

2.6 a

2.7 c

2.8 a

2.9 b

2.10 c

2.11 water droplets

2.12 clouds

2.13 more

2.14 rivers and streams

2.15 They all involve materials essential for life and they all require the atmosphere to complete the cycle.

2.16 oceans, streams, rivers, lakes, soil, plants, animals

2.17 Plants draw up moisture and evaporate (transpire) some of it. Some may seep into rivers. Some seeps through rock layers back to the sea.

2.18 Rain is formed by water vapor condensing on dust in the air. When the water droplets are large enough, they fall from the clouds.

2.19 Snow forms when water vapor condenses below the freezing point.

2.20 Some water is trapped in snow or ice for many years.

2.21 Example:
Kansas City receives 36 inches per year, more than the average

2.22 Student answer will read either increased or decreased.

2.23 If 2.22 was increased, the precipitation was greater. If 2.22 was decreased, the evaporation was greater

2.24 c

2.25 a

2.26 b

2.27 true

2.28 true

2.29 true

2.30 false

2.31 true

2.32 Either order:
a. photosynthesis
b. respiration

2.33 equals

2.34 a

2.35 c

2.36 Any order:
oceans, rocks, and living things

2.37 The oceans would absorb some of the carbon dioxide from the atmosphere.

2.38 The oceans would release some of the carbon dioxide to the atmosphere.

2.39 Photosynthesis changes carbon dioxide into oxygen and food containing carbon.

2.40 Respiration changes oxygen and food containing carbon into carbon dioxide.

2.41 The oxygen supply would run out in 2,000 years.

2.42 pressure and no oxygen

243 false

2.44 false

2.45 true

Science 705 Answer Key

2.46 true
2.47 false
2.48 Either order:
    a. plants
    b. bacteria
2.49 nitrogen compounds
2.50 Either order:
    a. soil
    b. roots of legumes
2.51 plant protein
2.52 ammonia
2.53 artificial fertilizer
2.54 Nitrogen is a component in many different compounds. It can release energy when changing from one compound to another.
2.55 lightning and blue-green algae
2.56 Bacteria use food provided by plants as energy to change nitrogen into nitrates.
2.57 He can plant legumes like clover, peas, or alfalfa and plow them into the soil at the end of the growing season.
2.58 They must eat plants or other animals.
2.59 Denitrification is the process by which nitrogen is released from nitrogen compounds by special bacteria.
2.60 Bacteria can release nitrogen from ammonia or change ammonia into nitrates.

SECTION THREE

3.1 a
3.2 c
3.3 b
3.4 a
3.5 d
3.6 c
3.7 b
3.8 a
3.9 a
3.10 waste substances, not ordinarily found in the atmosphere, that interfere with natural cycles
3.11 Either order:
    a. settling out
    b. rain or snow
3.12 carbon monoxide
3.13 ability of blood to transport oxygen
3.14 false
3.15 false
3.16 true
3.17 automobiles, electric power plants, factories, home furnaces, and burning dumps
3.18 particles and gases
3.19 Example:
The exhaust from cars could be cooled slowly to allow the nitrogen oxides to break apart.
3.20 a
3.21 c
3.22 c
3.23 a
3.24 lungs (or) nose
3.25 lungs
3.26 electric power plants and automobiles
3.27 life span or life expectancy

Science 705 Answer Key

3.28 Interferes with the blood's ability to carry oxygen. Dulls the senses, causing people to be prone to accidents.

3.29 smog and sulfur oxides

3.30 Damage to the lungs causes the heart to work harder.

3.31 warm air layer over a cool air layer can act like a lid holding pollution in the valley

3.32 b

3.33 c

3.34 d

3.35 a

3.36 dominion

3.37 rule or authority

3.38 man

3.39 false

3.40 false

3.41 Because God created the earth, He also knows the best way to enjoy the earth.

3.42 He is responsible for protecting his country and developing its resources.

3.43 The resources will be wasted and will run out. Pollution will become worse.

3.44 Matthew 7:12: The golden rule.

3.45 The Clean Air Act was passed. States have passed their own laws to control pollution.

3.46 It gives a reason for man to control pollution, by imposing penalties.

3.47 How much are we willing to pay to control pollution? What are we willing to do without to control pollution? How much pollution are we willing to accept?

3.48 teacher check

Science 705 Self Test Key

SELF TEST 1

1.01	e		1.013	b
1.02	c		1.014	a
1.03	a		1.015	c
1.04	d		1.016	b
1.05	f		1.017	false
1.06	c		1.018	mesosphere
1.07	b		1.019	thermosphere
1.08	g		1.020	equals
1.09	a		1.021	The main factor that changes is the temperature.
1.010	c		1.022	The gases in the troposphere, stratosphere, and mesosphere are mixed uniformly.
1.011	b			
1.012	b		1.023	The ozonosphere absorbs most of the ultraviolet rays that reach earth from the sun.

SELF TEST 2

2.01	g		2.013	d
2.02	j		2.014	b
2.03	i		2.015	a
2.04	c		2.016	b
2.05	h		2.017	b
2.06	e		2.018	a
2.07	a		2.019	a
2.08	d		2.020	d
2.09	b		2.021	b
2.010	f		2.022	b
2.011	d		2.023	a
2.012	a		2.024	a

Science 705 Self Test Key

2.025	d	
2.026	true	
2.027	true	
2.028	true	
2.029	true	

2.030 carbon dioxide and energy
2.031 energy and carbon dioxide
2.032 denitrification
2.033 legumes
2.034 plants or other animals

SELF TEST 3

3.01 n
3.02 l
3.03 c
3.04 m
3.05 o
3.06 a
3.07 j
3.08 e
3.09 g
3.010 k
3.011 i

3.012 f

3.013 b

3.014 h

3.015 b

3.016 a

3.017 c

3.018 a

3.019 a

3.020 d

3.021 b

3.022 a
3.023 b

3.024 d
3.025 b
3.026 a
3.027 a
3.028 b
3.029 a
3.030 c
3.031 c
3.032 b
3.033 b
3.034 b
3.035 true

3.036 false

3.037 true

3.038 true

3.039 true

3.040 stratosphere

3.041 denitrification

3.042 Either order:
    a. particles
    b. gases

3.043 ionosphere

3.044 Clean Air Act

Science 705
LIFEPAC Test

1. b
2. d
3. h
4. i
5. g
6. c
7. e
8. j
9. a
10. f
11. a
12. c
13. d
14. b
15. a
16. d
17. b
18. a
19. a
20. c
21. a
22. a
23. c
24. equals
25. plants or photosynthesis
26. oxygen
27. dominion
28. Golden Rule
29. false
30. true
31. true

Science 705 Alternate Test

Name _____

Match these items (each answer, 2 points).
1. ____ precipitation
2. ____ bacteria
3. ____ ozonosphere
4. ____ thermosphere
5. ____ troposphere
6. ____ ozone
7. ____ nitrogen
8. ____ oxygen
9. ____ automobile
10. ____ lungs

a. sphere of overturning
b. most common gas in the atmosphere
c. uncommon form of oxygen
d. where pollutants enter body
e. pollution
f. snow and rain
g. gas necessary for all living things
h. major source of pollution
i. hottest layer of the atmosphere
j. screens out ultraviolet rays
k. necessary organism in nitrogen cycle

Complete these statements (each answer, 3 points).
11. Matthew 7:12 is sometimes called the _____ .
12. Two types of air pollutants are a. _____ and b. _____ .
13. Bacteria and plants are necessary to complete the _____ cycle.
14. The main thing changing from layer to layer in the atmosphere is the _____ .
15. In order for plants to carry on photosynthesis, they must have _____ .
16. The probable reason for the increase of carbon dioxide in the atmosphere is _____ .
17. The two processes of life involved in the carbon-oxygen cycle are a. _____ and b. _____ .
18. The amount of carbon dioxide in the atmosphere is controlled by an exchange with the _____ .

Write the letter for the correct answer on each line (each answer, 2 points).
19. The process by which bacteria change nitrogen into nitrogen compounds is _____ .
    a. respiration   c. fixation
    b. denitrification   d. photosynthesis
20. The amount of radiation absorbed by the earth _____ the heat lost to space.
    a. is greater than   c. is less than
    b. equals   d. reduces

45

21. Carbon monoxide interferes with the transportation of _____ by the blood.
    a. oxygen   c. nitrogen
    b. carbon dioxide   d. hydrogen
22. Sulfur oxides harm the body by _____ .
    a. damaging the lungs
    b. allowing us to catch the flu
    c. keeping the body from making blood
    d. keeping the blood from carrying oxygen
23. In Genesis 1:26 God gave man the responsibility to have _____ over the earth.
    a. pollution   c. ownership
    b. dominion   d. energy

Answer *true* or *false* (each answer, 1 point).
24. ____ The process of changing a gas into a liquid is called evaporation.
25. ____ If man tries hard enough, he can eliminate all pollution.
26. ____ The rate at which oxygen is taken from the atmosphere by living things is equal to the rate at which plants replace it.
27. ____ Clouds reflect 65 percent of the incoming solar radiation.
28. ____ Steel corrodes two to four times faster in cities with many factories polluting the air.
29. ____ Man was given the responsibility to have authority over the entire earth.
30. ____ The trapping of heat by the earth's atmosphere is called the greenhouse effect.

54 / 67

Date _____

Score _____

Science 705 Alternate Test Key

1. f
2. k
3. j
4. i
5. a
6. c
7. b
8. g
9. h
10. d
11. golden rule
12. Either order:
    a. particles
    b. gases
13. nitrogen
14. temperature
15. carbon dioxide
16. burning fossil fuels
17. Either order:
    a. respiration
    b. photosynthesis
18. oceans
19. c
20. b
21. a
22. a
23. b
24. false
25. false
26. true
27. false
28. true
29. true
30. true

Science 707 Teacher Notes

Materials Needed for LIFEPAC
    Required:
    Atlas or online maps

    Suggested:
    reference materials in either book or online formats

Additional Learning Activities

  Section 1: Climate
    1. Use information from an encyclopedia, online resources, or almanac to show how average temperatures vary within the United States.
    2. Construct a bulletin board depicting the four elements of a climate.
    3. Make a map of the United States showing the average temperatures for a city in each state.
    4. Write a report on the causes of the wind, air pressure, precipitation, and temperature.
    5. Write a report on this topic: "Which affects me more, the weather or the climate?" (In which do you notice more changes?)

  Section 2: Climate: World Wide
    1. Make a large map of the world and label the general climactic areas.
    2. Make a map with isotherms or isobars. Give the map to a friend and ask him to interpret it.
    3. Use an encyclopedia or online resources to find the altitudes of the following cities or geographical locations: Death Valley, California; Innsbruck, Austria; Santiago, Chile; Beijing, China; Cape Town, South Africa; and Denver, Colorado. Make a chart to show the altitudes and mean annual temperatures of each place.

  Section 3: Climate: Regional
    1. Consult a daily newspaper or weather website and keep a weather chart on five cities for one week. Select one city in a desert climate, one in a steppes climate, one in a mountain climate, one in a mediterranean climate, and one in a humid subtropical climate.
    2. Cut pictures from magazines or print out images found online to illustrate the types of animals and vegetation you would find in a specific climate. Mount the pictures on poster board and label each item.
    3. Refer to an encyclopedia, online resources, or other library book to learn what life is like on a banana plantation. Write a one-page report on the information you find.
    4. Use an encyclopedia or online resources to study different types of housing and draw examples of each. Display your drawings in the classroom.

  Section 4: Climate: Local
    1. Read about the aborigines of Australia and write a report on them and their climate.
    2. Choose some places you would like to visit and find out what type of climates they have. Write a short paragraph about each place.

Science 707 Answer Key

SECTION ONE

1.1 the amount of heat or cold as measured in degrees
1.2 air pressure as measured by a barometer
1.3 a deposit of rain, sleet, snow ice, or hail
1.4 air in motion
1.5 short
1.6 a. know what to do
    b. know what to wear
1.7 false
1.8 false
1.9 a. cooler
1.10 a. climate
1.11 b. weather
1.12 Any order:
    a. temperature
    b. pressure
    c. wind
    d. precipitation
1.13 radiation
1.14 earth
1.15 more
1.16 14.7 pounds per square inch
1.17 heated (or warmed)
1.18 high
1.19 high
1.20 axis
1.21 c
1.22 f
1.23 b
1.24 e
1.25 d
1.26 Hailstones are formed when tiny ice particles move up and down in a windy cloud gathering more and more ice.
1.27 The rotation of the earth turns the winds to the right in the Northern Hemisphere and to the left in the Southern Hemisphere.

SECTION TWO

2.1–2.6 Any order:
2.1 latitude
2.2 altitude
2.3 nearness to water
2.4 average temperature
2.5 average rainfall
2.6 ocean currents
2.7 horse latitudes
2.8 doldrums
2.9 horse latitudes
2.10 Since the air is heated by the earth, the farther away it is, the less heat it receives.
2.11 Mountains cause clouds to rise, condense, and drop their rain on west slopes. No rain is left to fall east of the mountains.
2.12 Air over land is warmer than air over water. Warm air always rises. When this happens, cooler air from the ocean flows inland.

Science 707 Answer Key

2.13 The Gulf Stream warms them.
2.14 e
2.15 d
2.16 b
2.17 c
2.18 false
2.19 false
2.20 true
2.21 true
2.22 false
2.23 e
2.24 a
2.25 b
2.26 d
2.27 e
2.28 b
2.29 a
2.30 b

SECTION THREE

3.1 64.4
3.2 rain forests
3.3 jungles
3.4 furniture
3.5 Either order:
  a. rubber
  b. quinine
  or gum, furniture
3.6 summer
3.7 a seasonal wind reversal that brings heavy rainfall
3.8 a large farm
3.9 Examples; either order:
  a. coffee
  b. bananas
3.10 It is less constant because the savanna has a dry season each year.
3.11 true
3.12 false
3.13 false
3.14 farther from
3.15 dry
3.16 Urban
3.17 Mediterranean
3.18 They occur on the west coast and in the interior of continents on east side of mountain ranges.
3.19 The parched soil is too sun-baked to soak up the moisture, and it runs off.
3.20 It is a fertile spot in the desert where water can be found.
3.21 Nomads usually live in desert areas.
3.22 false
3.23 true
3.24 false
3.25 true
3.26 vertical
3.27 rain fall
3.28 They cause clouds to rise, condense, and drop their moisture
3.29 The topsoil and abundant rainfall make farming profitable.
3.30 The air is too thin and cold to support large plant life.
3.31 false

Science 707 Answer Key

3.32	true	3.49	a. humid subtropical
			b. cool
3.33	true	3.50	Any order:
			a. Switzerland
3.34	f		b. France
			c. Germany
3.35	a		d. Austria
3.36	g		e. Italy
3.37	b	3.51	polar
3.38	h	3.52	Sahara Desert
3.39	d	3.53	tropical rain forest and savanna
3.40	e	3.54	true
3.41	generally near the poles	3.55	false
3.42	seven	3.56	true
3.43	Asia	3.57	false
3.44	Siberia	3.58	true
3.45	steppe	3.59	false
3.46	Gobi Desert	3.60	false
3.47	Himalayas	3.61	true
3.48	Mediterranean	3.62	false

SECTION FOUR

4.1	M	4.11	Agriculture is more easily carried on. Machines do much of the work. Jobs take only 8–10 hours per day.
4.2	L		
4.3	M	4.12	The cycles of plowing, sowing, and harvesting control life.
4.4	H		
4.5	H	4.13	The making of an area into a city and a people into city people.
4.6	H	4.14	Few people in these areas can afford more substantial homes. Permafrost prevents the building of brick or concrete homes in most cases.
4.7	M		
4.8	H	4.15	The warmth of the low latitudes makes warm clothing unnecessary; in the high latitudes warm clothing is essential for protection from the weather.
4.9	L		
4.10	M		

Science 707 Answer Key

4.16	true		4.28	b. nomads
4.17	false		4.29	b. December and January
4.18	false		4.30	c. at the oasis
4.19	true		4.31	It provides food, shelter, clothing, transportation, and utensils.
4.20	true		4.32	They would like to steal valuable camels.
4.21	true		4.33	It is thatched, has one door, no windows, and one hole in the roof.
4.22	true		4.34	Any order:
4.23	false			a. bows and poisoned arrows
4.24	false			b. fish hooks
4.25	b. harvest			c. spears
4.26	a. sometimes		4.35	lack of berries and roots to collect
4.27	a. Entertainment		4.36	teacher check

Science 707 Self Test Key

SELF TEST 1

1.01	false		1.015	weather
1.02	false		1.016	climate
1.03	false		1.017	solar radiation
1.04	true		1.018	water
1.05	true		1.019	trade winds
1.06	true		1.020	a. west
1.07	true			b. east
1.08	true		1.021	polar easterlies
1.09	true		1.022	prevailing westerlies
1.010	false		1.023	northeast trade winds
1.011	14.7 lbs. per sq. inch		1.024	southeast trade winds
1.012	hail (or hailstones)		1.025	prevailing westerlies
1.013	rotates		1.026	polar easterlies
1.014	equator			

SELF TEST 2

2.01	true		2.012	most
2.02	true		2.013	cold
2.03	true		2.014	lower
2.04	false		2.015	humidity (rainfall or heat)
2.05	false		2.016	breeze
2.06	true		2.017	maritime
2.07	false			
2.08	true			
2.09	true			
2.010	true			
2.011	Example:			

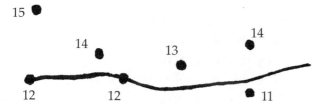

Science 707 Self Test Key

2.018 equator
2.019 horse latitudes
2.020 equator
2.021 Labrador Current

2.022 isotherms
2.023 west
2.024 axis

SELF TEST 3

3.01 d
3.02 j
3.03 g
3.04 l
3.05 b
3.06 e
3.07 i
3.08 a
3.09 h
3.010 c
3.011 equator
3.012 savannas
3.013 climate

3.014 from which they blow
3.015 dry
3.016 Either order:
    a. North Pole
    b. South Pole
3.017 permafrost
3.018 Antarctica
3.019 Gulf Stream
3.020 cold
3.021 60° N
3.022 30° N
3.023 0°
3.024 30° S
3.025 60° S

SELF TEST 4

4.01 camel
4.02 wooden
4.03 oasis
4.04 market
4.05 Pygmies
4.06 Bedouins
4.07 thatched
4.08 seals
4.09 poisoned
4.010 winter

4.011 polar easterlies
4.012 prevailing westerlies
4.013 northeast tradewinds
4.014 southeast tradewinds
4.015 prevailing westerlies
4.016 polar easterlies
4.017 tropical rain forest
4.018 tropical savanna
4.019 the Gulf Stream
4.020 because of permafrost
4.021 desert (arid)
4.022 trading
4.023 hunting and gathering
4.024 answer will fit community where project is completed
4.025 furs
4.026 farming

Science 707
LIFEPAC Test

1. true
2. true
3. true
4. false
5. true
6. true
7. false
8. false
9. false
10. true
11. i
12. k
13. h
14. j
15. a
16. c
17. e
18. b
19. d
20. g
21. c. fur trading
22. a. poisoned arrows
23. a. farming
24. c. Labrador Current
25. a. very warm
26. b. hail
27. a. dry
28. a. 0°
29. c. the Sahara Desert
30. c. meteorologists

Science 707 Alternate Test

Name _____

Match these items (each answer, 2 points)
1. _____ jungle
2. _____ Gobi
3. _____ nomadic
4. _____ steppes
5. _____ Mediterranean climate
6. _____ Pygmy
7. _____ doldrums
8. _____ tree line
9. _____ Australia
10. _____ permafrost

a. an African of small stature
b. ocean regions near the equator
c. wandering desert tribesman
d. deeply frozen earth
e. wet winters, dry summers
f. urban areas
g. plains
h. a large desert in Asia
i. a rain forest of small trees, bushes, and vines
j. its center is a large, desert area
k. the point above which trees do not grow

Write the letter for the correct choice on each line (each answer, 2 points).

11. Precipitation composed of lumps of ice is called _____.
    a. hail   b. sleet   c. snow
12. Pygmies use primitive weapons such as _____.
    a. muskets   b. harpoons   c. poisoned arrows
13. The Eskimos earn money mainly by _____.
    a. flying   b. fishing   c. fur trading
14. The population of a steppe community is mostly engaged in _____.
    a. governing   b. farming   c. trading
15. The climate to the east of a mountain range is usually very _____.
    a. dry   b. cold   c. wet
16. Temperatures near the equator are usually _____.
    a. somewhat warm b. very warm   c. freezing
17. Almost all of northern Africa is occupied by _____.
    a. the Sahara Desert b. the Great Oasis c. permafrost
18. The equator is a latitude of _____.
    a. 0°   b. 60°   c. 90°
19. The _____ is an ocean current bringing cool air from the north.
    a. North Atlantic Drift b. Labrador Current c. Gulf Stream
20. The condition of the air over a region averaged through a period of years is _____.
    a. weather   b. climate   c. air pressure

Science 707 Alternate Test

Answer *true* or *false* (each answer, 1 point)
21. _____ Antarctica has a polar climate.
22. _____ Above the Arctic Circle the sun is almost directly overhead.
23. _____ Climate is the day-to-day changes in the air.
24. _____ Normal sea level air pressure is 14.7 pounds per square foot.
25. _____ The camel is the mainstay of most Bedouins.
26. _____ Eskimos live mainly in wooden shacks.
27. _____ Winds are turned to the right in the Northern Hemisphere.
28. _____ Mountainous areas have little or no rainfall.
29. _____ Places having equal temperatures are joined by isotherms on weather maps.
30. _____ In the humid subtropics the temperatures can drop below freezing.

Complete these activities (each numbered item, 5 points).
31. List the four parts of climate.
    a. _____
    b. _____
    c. _____
    d. _____
32. Name the type of climate you live in and explain your answer.
    _____
    _____
    _____

48 / 60

Date _____

Score _____

Science 707 Alternate Test Key

1. i
2. h
3. c
4. g
5. e
6. a
7. b
8. k
9. j
10. d

11. a. hail
12. c. poisoned arrows
13. c. fur trading
14. b. farming
15. a. dry
16. b. very warm
17. a. the Sahara Desert
18. a. 0°
19. b. Labrador Current
20. b. climate
21. true
22. false
23. false
24. false
25. true
26. true
27. true
28. false
29. true
30. true
31. Any order:
    a. temperature
    b. pressure
    c. wind
    d. precipitation
32. teacher check: Responses of students will depend on their geographical location.